innovative timber construction

new ways to achieve energy efficiency

Robin Lancashire and Lewis Taylor

TRADA Technology Ltd
Chiltern House
Stocking Lane
Hughenden Valley
High Wycombe
Buckinghamshire HP14 4ND

tel: +44 (0)1494 569600
fax: +44 (0)1494 565487
email: information@trada.co.uk
website: www.trada.co.uk

ISBN 978-1-900510-87-5

Published in 2012 by TRADA Technology Ltd, reprinted 2014

TRADA Technology acknowledges the assistance of Roddy Langmuir of Edward Cullinan Architects and Patrick Hislop RIBA and for their review of an early draft of the text and illustrations

Main cover photograph: Engineered stud walls in St Luke's C of E Primary School, Wolverhampton. Architect: Architype. Photo: Leigh Simpson

innovative timber construction
new ways to achieve energy efficiency

As thermal performance requirements have become harder and more expensive to achieve, all sectors of the construction industry need to consider building a thicker external envelope to provide space for insulation. This need is only going to increase as the industry moves toward the 'Zero Carbon' target in 2016. For timber frame construction, merely increasing the depth of solid timber studs and fitting more insulation between them is becoming insufficient to meet the increasing requirements for thermal and energy performance.

Although enthusiasts have designed a variety of low-energy timber frame buildings over recent decades, the techniques used have remained on the periphery of the construction industry, with mainstream timber frame evolving slowly to keep pace with national building regulation changes. This book has been written to expand on information provided on some alternative forms of timber construction mentioned in *Timber Frame Construction 5th Edition*.

Four of the most common types of alternative timber walls and roofs are explored here. These are structural insulated panels (SIPs), cross-laminated timber (crosslam), engineered stud and twin stud.

Robin Lancashire is head of the Building Performance section at TRADA Technology with particular expertise in timber frame construction.

Lewis Taylor is a timber frame consultant at TRADA Technology with particular expertise in thermal and acoustic performance.

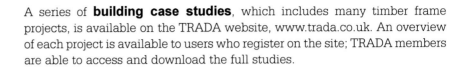

A series of **building case studies**, which includes many timber frame projects, is available on the TRADA website, www.trada.co.uk. An overview of each project is available to users who register on the site; TRADA members are able to access and download the full studies.

TRADA, the Timber Research and Development Association, is a not-for-profit, membership-based organisation delivering key services to members in support of its two main aims of 'Building markets for timber' and 'Increasing specification'. Membership encompasses companies and individuals across the entire timber supply/use chain, from foresters and sawmillers, through merchants and manufacturers, to architects, engineers and specifiers.

For further information and details of membership visit www.trada.co.uk or telephone 01494 569603.

TRADA Technology is an independent consultancy company providing a wide range of commercial and training services to the timber and construction industries. Prior to 1994 it was wholly owned by TRADA, the Timber Research and Development Association. It is now a member of the BM TRADA Group of companies and is TRADA's appointed provider for its research and information programmes and for the administration of its membership services.

For online technical information, consultancy services and training visit www.tradatechnology.co.uk or telephone 01494 569600.

Contents

1 Introduction

As thermal performance requirements have become harder and more expensive to achieve, all sectors of the construction industry have been driven down the path of building a thicker external envelope to provide space for insulation. For timber frame construction it has now reached a stage when merely increasing the depth of solid timber studs and fitting more insulation between them is becoming insufficient. To keep up to date with current thermal performance regulations, the details in *Timber Frame Construction 5th Edition*[1] have introduced an insulated service zone on the inside of the wall to help improve airtightness and allow space for more insulation to be installed. Depending on the insulation materials used, U-values below $0.2\,\mathrm{W/m^2K}$ can be achieved.

Timber Frame Construction predominantly deals with what could be considered 'conventional' platform timber frame construction – the panelised form of construction using solid timber wall studs and rails with wood-based sheathing boards. This method has been embraced by the UK construction industry as a standard form of timber frame construction over the last 50 years.

Although enthusiasts have designed a variety of low-energy timber frame buildings over recent decades, the techniques used have remained on the periphery of the construction industry, with mainstream timber frame evolving slowly to keep pace with national building regulation changes. Driven by the need to reduce CO_2 emissions, the European Union has started to set ever-tighter requirements on the energy performance of buildings, leading to 'zero-carbon' buildings by 2016.

This book has been written to expand on information provided on alternative forms of timber construction mentioned in *Timber Frame Construction*. Four of the most common types of alternative timber walls and roofs are explored here. These are structural insulated panels (SIPs), cross-laminated timber (crosslam), engineered stud and twin stud. Although these systems are all different, they still follow the basic principles of timber frame construction that are set out in *Timber Frame Construction*. Although this book has been written as a standalone publication, it is not intended to be a full technical manual for each construction type, but demonstrates what is possible and how each of the forms works in principle to provide an effective thermal envelope.

More industry changes will be needed to meet 2016 regulations. In order to achieve further reductions in U-values and achieve buildings with minimal energy requirements, timber frame designers and architects are beginning to explore some of these alternative types of timber frame construction, which have been on the periphery for many years but are now becoming better appreciated for their high levels of thermal performance.

1.1 Design requirements

Timber frame buildings have demonstrated that they are a durable form of construction and can easily achieve a minimum design life of 60 years. There are many examples of softwood timber frame buildings that are hundreds of years old. In addition, timber frame buildings have evolved to provide good thermal and acoustic performance. It is important to remember, when making further improvements in thermal performance, to maintain the same good design principles.

1.1.1 Designing for durability

The single largest factor in the durability of timber is exposure to moisture. The decay threshold for timber is a moisture content (MC) in excess of 20%. If timber remains at an MC of 20% or less, the timber will not be subject to mould growth or decay. In practice, the MC of structural timbers in a well-designed and constructed timber frame external wall will be nearer 12%, well below the decay threshold. In order to ensure that the wood in a timber frame external wall remains at a low MC, a number of layers of protection are built into the wall. Typically the timber frame wall panels have a vapour control layer (VCL) internally, are lined with a breather membrane outside the timber frame and are encapsulated in a cladding system, incorporating a drained and ventilated cavity between the cladding and the timber frame. Alternative forms of timber construction should follow these principles too.

A breather membrane is a material which has a low moisture vapour resistance, but the ability to resist liquid water penetration. During construction of the building, the breather membrane helps to provide a level of protection to the exposed wall panels. Once the building is complete, if moisture penetrates the cladding and cavity (for example, wind-driven rain penetration) the breather membrane deflects moisture away from the timber frame. The low moisture vapour resistance of the membrane ensures that any moisture vapour trapped within the wall is able to escape out into the external wall cavity.

The cladding system, whether constructed of masonry or a lightweight system, is designed to provide a rainscreen function, keeping the majority of moisture away from the structure. To ensure that any moisture which does penetrate the cladding does not affect the timber structure, a drained and vented cavity is incorporated behind the cladding. This cavity can range from 20mm for timber cladding up to 50mm for masonry cladding and will incorporate drainage and ventilation gaps at the base. *BS 5250:2002 Code of practice for control of condensation in dwellings*[2] states that vented air spaces should have 'openings to the outside air of not less than 500mm² per meter length of wall'. This requirement should be repeated at each floor level if horizontal cavity barriers or cavity trays are installed in the floor zone.

VCLs are used to manage the movement of moisture vapour from the warm to the cold side of a wall (such as inside to outside in winter). It is important to ensure that the layers on the cold side of a wall are more vapour permeable than those on the warm side. This ensures that as moisture vapour passes through the wall, interstitial condensation does not occur. As a general rule of thumb, the VCL on the warm side of the wall should have a moisture vapour resistance five times greater than the layers on the cold side of the wall.

It is worth remembering that a VCL does not have to be a sheet of plastic. In many situations, where a wood-based structural sheathing board is being used on the outside of the timber frame, the VCL needs to have a very high moisture vapour resistance (and typically a polythene membrane is used). However, if that structural sheathing board is being used on the inside of the wall, the sheathing board itself may be acting as the VCL and no additional membranes are required. In cross-laminated timber (crosslam) structures, where all the insulation is placed on the outside of the timber wall panels, the crosslam structure itself is performing a vapour control function.

Therefore, detailing at junctions and interfaces should be robust in design and construction to achieve this. Interstitial condensation risk calculations must always be carried out to ensure that no risk of condensation is present.

With all timber-based building methods, it is important to keep the base of the building above external ground level to mitigate the risk of ground water coming into contact with the structural timber. It is recommended that the lowest structural timber (usually the sole plate) is at least 150mm above external ground level. This is also a requirement of most warranty providers.

Ensuring that the base of the timber frame is above external ground level is normally relatively straightforward; however, the provision of level access into the building can sometimes cause conflict between ground level and the lowest structural timber. In these cases it is normal to provide an upstand of masonry or concrete to lift the timber up away from external ground level.

1.1.2 Acoustic performance

Requirements to demonstrate the sound insulation performance of external walls are normally based on the location of the building, the building type and any nearby noise sources, such as a major road or rail line. If requirements are imposed to provide specific external wall sound insulation, the performance of the facade will normally be dictated by the ambient background noise at the site and target internal noise requirements. Typically, the sound transmission through an external wall is dominated by the performance of any windows rather than the performance of the external wall itself. If a higher performance is required (due to very high external noise levels and very high-performance windows), multiple layers of acoustic-rated plasterboard will normally provide a sufficient improvement.

Internal walls and party walls in timber frame buildings (regardless of specific external wall construction type) are typically constructed from timber studwork and lined with plasterboard. These walls have a proven track record of good acoustic performance and compliance with national building regulations. SIPs or crosslam may be used for internal and party walls, as well as external walls. Typically the SIP or crosslam manufacturer will have acoustic performance data, based on either site or laboratory test evidence, to justify the performance of their system and demonstrate compliance with regulations.

1.1.3 Fire performance

All forms of construction need to comply with the national building regulations' fire performance requirements and there is no difficulty in meeting the required levels with alternative timber-based structures, given correct design, standards of manufacture and workmanship.

The fire resistance of a completed timber frame structure is achieved by a combination of the internal lining material, the timber structure and insulation. Fire-resistance requirements for elements of construction are defined within national building regulations.

If timber is protected from direct attack by a fire source (such as in a timber frame building with plasterboard linings), it cannot ignite and burn before a temperature in excess of 400°C is reached at the timber surface. Contrary to some people's perceptions, timber used in construction performs well against fire. Timber burns steadily at a predictable rate; and in the process, charcoal is formed on the surface of the timber, which serves to insulate and protect the core. When fire does take hold, timber reacts differently to other common structural materials, with the following advantages:

- uniform charring at a low rate (after the protective plasterboard has fallen away)
- low heat conduction
- no deformation at high temperatures.

SIPs, engineered stud and twin stud walls will all use plasterboard to provide the required period of fire resistance to the structure. Due to their small timber section sizes, these three forms provide limited contribution to the overall performance of the structure in the event of a fire. Crosslam buildings, on the other hand, can rely heavily on charring of the wall panels to provide some or all of the required fire resistance. Char rate calculations, based on the information provided in *Eurocode 5*[3] would need to be undertaken and a structural engineer would need to ensure that there is sufficient redundant material in the wall panel to allow this level of charring to the structure. If the crosslam structure is to be left exposed, surface spread of flame performance must be considered.

1.2 Differential movement

All buildings constructed from loadbearing timber components will be subject to differential movement. Differential movement occurs in a newly completed timber building when the structural timber components dry and the building settles. Differential movement only occurs once, taking around two years before movement is complete (subject to occupancy patterns and building type). Allowances for differential movement should therefore be made in all timber frame buildings. Detailed guidance on differential movement can be found in *Timber Frame Construction*.

1.3 Material specifications

Detailed guidance on the properties and specifications of common construction materials can be found in Appendix 1 of *Timber Frame Construction*.

1.4 Thermal performance

'Fabric first' is a popular ethos that has grown from the German Passivhaus standard. This is where energy-efficient buildings are created by focusing on the performance of the external envelope of the building, before looking to renewable energy sources or 'bolt-on' technologies. With the fabric-first approach, elemental U-values and heat loss are driven down to very low levels, so the building consumes a minimal amount of energy, staying warm in the winter and cool in the summer. This is the aim of the alternative timber external envelopes discussed in this book.

Energy efficiency of buildings is generally broken down into three key areas of performance:

- U-values of building elements
- Ψ-values (thermal bridging) of element junctions
- air permeability of the building envelope.

U-values of building elements refer to the heat loss through the walls, floors, roof, windows and external doors. National building regulations specify minimum U-value performance criteria for building elements. In reality, the performance required for walls, floors and roofs is normally much better. The constructions shown in this book can achieve superior U-values because they are well insulated. For all but crosslam, because the insulation is part of its structure, there is a savings in space which helps to keep the envelope slender.

Thermal bridging (Ψ-value) occurs in all construction types and is caused by areas of reduced insulation or where an element passes through the insulation. Timber has a lower thermal resistance than the insulation materials placed between the framing members. Therefore, greater heat flow occurs through studs, plates, rails and joists than in other areas of the external wall or roof structures. This increase in thermal conductivity is known as thermal bridging. In general, thermal bridges can occur at any junction between building elements or where the building structure changes. The forms shown in this book aim to reduce thermal bridging to a minimum. There are two categories of thermal bridges, repeating and non-repeating.

Repeating thermal bridges

The additional heat flow resulting from repeating thermal bridges is included in the calculation of a particular building element U-value that contains these thermal bridges, such as studs in timber frame walls.

Non-repeating thermal bridges

The additional heat flow resulting from this type of thermal bridge is determined separately, either by the numerical calculation method given in *BS EN ISO 10211-1*[4] or by computer modelling using finite element modelling software, and is known as a Ψ-value.

Recent changes to national building regulations now require that non-repeating thermal bridging is considered when calculating the energy consumption and CO_2 emissions from buildings. Ψ-values are specific to each building system, as well as to each individual building, and so should be calculated on a project-specific basis.

1.5 Air permeability

Air permeability is another major aspect of energy efficiency. National building regulations require that unplanned air leakage into, or out of, the building is controlled. Airtightness tests are normally carried out once the building is virtually complete, and targets for airtightness are normally set based on the Standard Assessment Procedure (SAP) calculations.

It is important to understand that buildings with very low building-fabric air-leakage rates will require additional ventilation measures. Typically these will take the form of mechanical ventilation heat recovery (MVHR) units. MVHR units mechanically control the movement of air into and out of a building while recovering latent heat in the exhaust air.

1.6 Walls

Alternative timber wall panels can be as thick as necessary to achieve the required thermal performance. Many new-build developments will have external wall U-value targets of between 0.1 W/m²K and 0.2 W/m²K, which all the alternative wall types can easily achieve.

External walls will generally consist of the following layers:

- external cladding
- drained and ventilated cavity
- breather membrane
- alternative wall and insulation
- VCL
- battens forming a service void (possibly insulated)
- internal linings.

The overall thickness of a wall will be dictated by its thermal performance, rather than by structural or other requirements. Reflective breather membranes and VCLs can also provide a worthwhile contribution to the overall U-value of the external wall. Reflective membranes, coupled with an adjoining air gap, create a low-emissivity void that can enhance U-values by as much as $0.02\,W/m^2K$.

If a service void is used on the inner face of the external wall, it is important to ensure that external air does not enter this void. If external air passes into/through this void, convection air movement can reduce the thermal performance of the wall. This is known as thermal bypass. To ensure that this does not occur, it is important to make sure that the air barrier is well sealed and continuous, including around any service penetrations. If possible, penetrations to the air barrier should be designed out. To further reduce the risk of convection air movement, insulation can be installed within the void. It is recommended that mineral fibre insulation is used because it can be easily fitted around services and, if firmly compressed, will all but eliminate convection air movement in the void.

1.7 Roofs

Both pitched and flat roofs can be formed using materials similar to those used for walls. It should be remembered that when the structure of a roof passes through the insulation, it is subject to a temperature gradient and is therefore defined as a cold roof. Cold roofs, whether pitched or flat, will require a ventilation void between the roof structure and the roof covering.

Many new-build developments will have roof U-value targets of between $0.1\,W/m^2K$ and $0.15\,W/m^2K$.

Pitched cold roofs will generally consist of the following layers:

- tiles or slates
- tiling battens
- counter battens (providing a path for ventilation and water runoff)
- breathable roofing membrane
- roof structure and insulation
- VCL
- battens forming a service void (possibly insulated)
- internal lining.

National building regulations and British Standards should be consulted for the size of the ventilation voids required in flat and pitched roofs. These regulations and standards may also pose limits on the size/spans of flat roofs for ventilation requirements.

2 SIPs

A structural insulated panel (SIP) consists of a layer of oriented strand board (OSB) bonded onto each side of an insulating foam core. A strong, structural bond between the three layers is essential to the loadbearing ability of the SIP. The resultant sandwich of OSB and foam can be used as a structural loadbearing element. A SIP is a composite engineered product, with each building being individually designed and engineered. SIP structures can offer good thermal and airtightness performance with minimal thermal bridging due to the continuous nature of the insulation core within the panel.

SIP buildings were first researched in North America in the 1930s and 40s. SIPs as we know them today were conceived in the 1970s and have been in use since then.

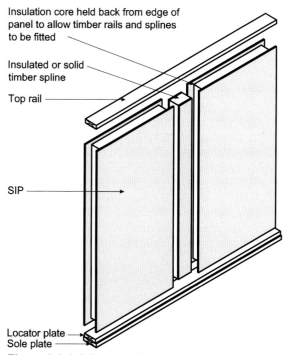

Insulation core held back from edge of panel to allow timber rails and splines to be fitted

Insulated or solid timber spline

Top rail

SIP

Locator plate
Sole plate

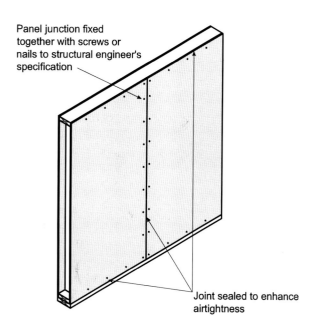

Panel junction fixed together with screws or nails to structural engineer's specification

Joint sealed to enhance airtightness

Figure 2.1 Joining panels

2.1 Applications

SIPs are used for external wall and roof structures where good thermal performance with minimal thermal bridging is desired. Currently, load-bearing SIP structures are generally limited to four storeys in height for structural reasons. As SIP structures become larger or more complex, additional elements of structure may be required to support the imposed loads; however, these elements of structure will reduce the thermal performance benefits of the SIP. SIPs can also be used as in-fill panels on large multi-storey steel- or concrete-framed buildings. Large-format SIPs can be craned into place and fixed to the outside of the frame, providing rapid enclosure of the structure with thermal insulation already built into the walls. SIP walls can be clad with all typical cladding materials, such as masonry, timber or other lightweight systems.

SIPs can form both pitched- and flat-roof structures and can be finished with any form of normal roof covering. Pitched SIP roofs usually consist of large panels supported on purlins and ridge beams. This provides an open, unobstructed roof void that is part of the thermal envelope of the building and ideal for use as a room-in-roof structure.

SIPs are not normally used in the construction of floors, predominantly due to the structural performance of the panel when used horizontally. Typically, floor systems in SIP buildings consist of traditional timber joist floor structures.

2.2 Sustainability

Most SIPs manufactured in the UK will be constructed from OSB/3 made in the British Isles from home-grown timber. This helps to reduce the embodied energy (transport emissions) associated with the manufacture of engineered timber.

Most SIP cores are constructed from expanded polystyrene (EPS), polyurethane (PUR) or polyisocyanurate (PIR) phenolic foam insulation. Although these materials come from non-renewable sources, the energy efficiency over the lifetime of the finished SIP structure should help to offset the initial manufacture of the insulation.

2.3 Structure

SIP buildings are typically designed and built as platform-frame structures. In platform-frame structures, floors bear on to the head of the walls below, with subsequent wall panels installed bearing on top of the floor structure above the lower wall panels. This process is then repeated for each storey of the building. The structural calculations to support the design of a SIP building should be carried out in accordance with *Eurocode 5*.

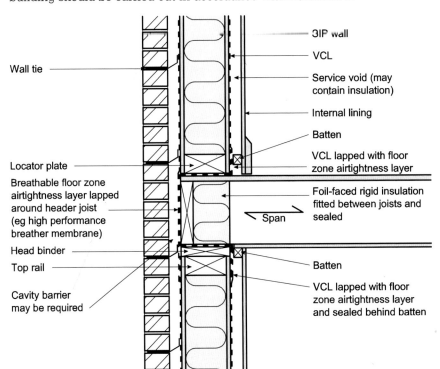

Figure 2.2 SIP floor
Gaps in construction shown to illustrate membrane laps.

Racking resistance of a SIP wall is typically greater than that of a timber studwork wall. Therefore, a low-rise building designed to be constructed with loadbearing timber studwork should also work as a loadbearing SIP structure. A structural engineer would need to confirm this by calculation.

Normally, the structural calculations will be undertaken by the SIP manufacturer or supplier. SIP manufacturers will have data relating to racking resistance, loadbearing capacity and flexural and bending strength. These values are based on test data and may vary between panel manufacturers.

Timber battens, cladding, wall linings, etc are normally fixed to the SIP using screws. The loadbearing capacity and pull-out resistance of screw fixings into the SIP can either be obtained from manufacturers' test data, or calculated in accordance with *Eurocode 5*.

2.4 Manufacture

Elements of building structure are manufactured as either small individual panels, typically 1.2m × 2.4m (standard OSB sheet-sized panels) or as larger panels made from multiple or single large-format sheets of OSB. Panels with a length or height up to around 6m can be manufactured to provide double height spaces or large panels for increased speed of construction.

There are two distinct ways in which the panels are produced. The first involves bonding the OSB sheathing board onto a block of pre-manufactured foam insulation. The second involves the injection of liquid expanding foam between two layers of board that are held apart in a jig at the desired panel thickness. As the foam expands, it fills the void between the boards and bonds to them at the same time.

Once the panels have been manufactured, openings in wall panels can be formed. These can either be in the form of small framing panels that are site-assembled to form the opening as shown in (a) and (b) in *Figure 2.3*, or cut from a large blank wall panel. Solid timber lintels, studs and rails are normally inserted around the perimeter of the opening into the core of the panel to support imposed loads from above. Alternatively, floor joists may be hung from a header joist over the opening to eliminate the need for lintels within the SIP (c). These timbers support any imposed loads from above (lintel, floor or roof loads) and provide a solid fixing for the installation of window or door sets. The location and size of window openings are considered in the site-specific structural calculations for racking resistance. If required, openings can be moved and additional openings formed with relative ease; however, the size, location and method of creating these openings must be designed by a structural engineer.

Openings for roof lights in roof panels are formed in the same way as similar openings in wall panels. As with walls, additional or new openings can be formed with relative ease under the guidance of the structural engineer.

2.5 Erection

Wall panels bear onto the foundations via a preservative-treated timber sole plate which is firmly secured to the sub-structure. The purpose of this horizontal timber sole plate is to provide a levelled substrate onto which the wall panel can be fixed and to transfer vertical load to the foundations. The way in which the wall panel bears onto the sole plate can differ between systems.

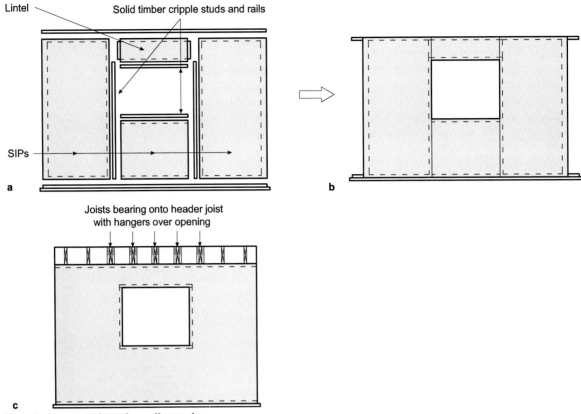

Figure 2.3 SIP window openings in wall panels

Some systems incorporate a solid timber rail into the base of the SIP, which in turn bears onto the sole plate. Other systems do not include this rail, and the inner and outer leaves of OSB fit over the sides of the sole plate. Either way, the wall panel is fixed to the sole plate to provide solid location of the panel and resistance to sliding, overturning and uplift forces. As with all timber-based buildings, all structural timber should be at least 150mm above external finished ground level to help ensure long-term durability.

Panels are joined using either solid timber within the ends of the panels or SIP-based insulated splines, depending on the manufacturer's specific system and/or structural requirements. The splines are fixed into the ends of the panels and sealed using a flexible mastic or an expanding foam glue. At corner junctions, long (200mm to 300mm+) specialist screw fixings are typically used to pass through the side of one panel into the end of the adjoining panel. Again, panels are sealed with flexible mastic or expanding foam glue.

Roof panels are normally joined to wall panels using a tapered timber fillet fixed to the head of the wall panel. The roof panel would then rest on the tapered fillet, as well as purlins and ridge beams, and be fixed into place with large screws which pass through the SIP into the timber element below.

2.6 Thermal performance

2.6.1 U-values
Walls
SIP wall panels can be of any thickness, although many manufacturers will make panels between 100mm and 250mm thick. Most new-build

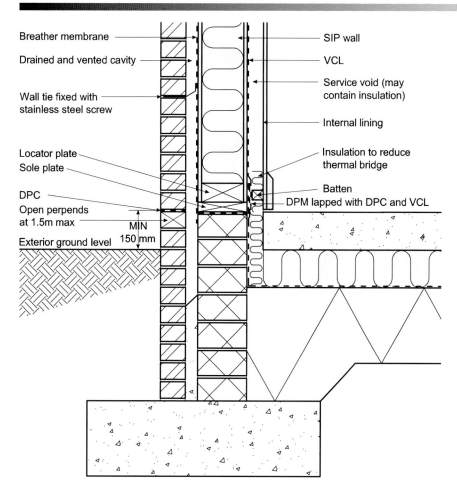

Breather membrane
Drained and vented cavity
Wall tie fixed with stainless steel screw
Locator plate
Sole plate
DPC
Open perpends at 1.5m max
Exterior ground level
MIN 150 mm

SIP wall
VCL
Service void (may contain insulation)
Internal lining
Insulation to reduce thermal bridge
Batten
DPM lapped with DPC and VCL

Figure 2.4 SIP foundation
Gaps in construction shown to illustrate membrane laps. All structural timber should be at least 150mm above external ground level.

developments will have external wall U-value targets of between 0.1 W/m²K and 0.2 W/m²K, which SIPs can easily achieve with these wall thicknesses.

External walls will generally consist of the following layers:

- external cladding
- drained and ventilated cavity
- breather membrane
- SIP of the required thickness
- VCL
- battens forming a service void (this void may contain additional insulation)
- internal linings.

The SIP itself provides the greatest contribution to the overall U-value of the external wall. In most instances, U-value targets will dictate the overall thickness of the SIP, rather than structural or other requirements. Reflective breather membranes and VCLs can also provide a worthwhile contribution to the overall U-value of the external wall when coupled with an adjoining air gap to create a low-emissivity air void.

The SIP manufacturer should be able to conduct U-value calculations based on their system and client specifications.

If U-values better than those achievable with the SIP alone are required, additional insulation could be installed to the inside or outside of the SIP. The thickness and type of insulation would need to be considered and condensation risk calculations conducted. A drained and vented cavity should always be maintained behind the external cladding (for example, insulated render systems should not be installed directly to the SIP with no cavity).

Roofs

Both pitched and flat roofs can be formed using SIPs. SIP roof panels can be of any thickness, although most manufacturers will make panels between 100mm and 250mm thick. Most new-build developments will have roof U-value targets of between $0.1\,W/m^2K$ and $0.15\,W/m^2K$, which SIPs can easily achieve.

Pitched SIP roofs will generally consist of the following layers:

- tiles or slates
- tiling battens
- counter battens (providing a path for ventilation and water runoff)
- breathable roofing membrane
- SIP of the required thickness
- VCL
- battens forming a service void (this void may contain additional insulation)
- internal lining.

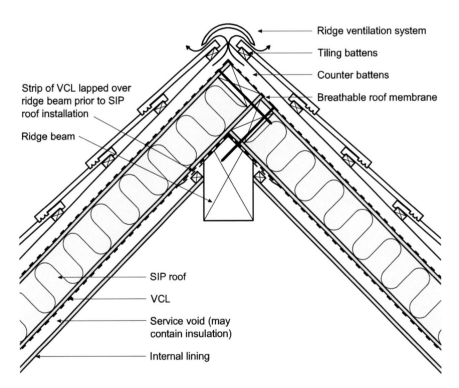

Figure 2.5 SIP ridge

Flat SIP roofs (*Figure 2.9*) will generally consist of the following layers:

- waterproof roofing membrane
- plywood or similar roof decking
- furrings to form a runoff and create a ventilation void
- breathable roofing membrane
- SIP of the required thickness
- VCL
- battens forming a service void (this void may contain additional insulation)
- internal lining.

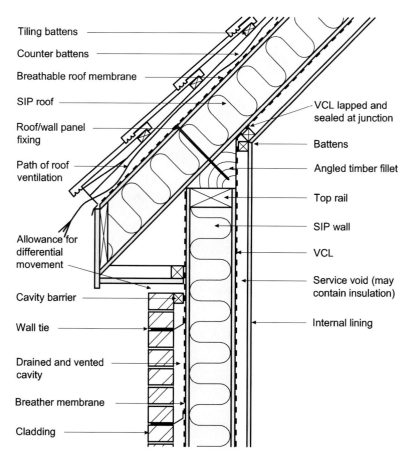

Figure 2.6 SIP wall/roof junction

The SIP manufacturer should be able to conduct U-value calculations based on their system and client specifications.

As with walls, if U-values better than those achievable with the SIP alone are required, additional insulation could be installed to the inside or outside of the SIP. The thickness and type of insulation would need to be considered, and condensation risk calculations conducted.

2.6.2 Ψ-values (thermal bridging)

Although SIPs are largely constructed from insulating foam and OSB sheathings, SIP walls and roofs contain solid timber around window and door openings as well as corner junctions, and may contain solid timber at panel junctions too. The proportion of solid timber within a SIP element will be entirely dependent on the building type and size and should be considered on a site-by-site basis. However, the timber content can be as low as 4%, versus an average of 15% for traditional timber frame buildings. The specific timber fraction should be calculated or obtained from the SIP manufacturer and used in the calculation of U-values.

SIPs can reduce non-repeat thermal bridging when compared to solid-timber studwork timber frame walls; however, most SIPs contain solid timber at panel junctions, which results in thermal bridging similar to that in traditional timber frame. It may be possible to reduce the amount of timber through structural calculation, or reduce the effects of the bridging with the inclusion of additional insulation in the service void formed on the inside of the wall.

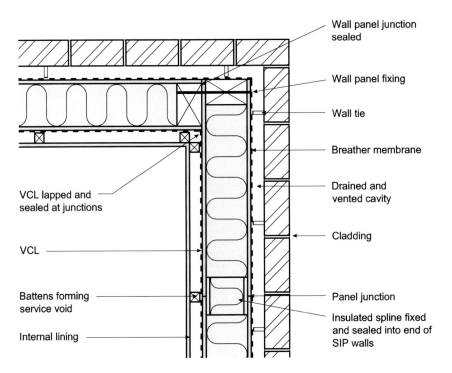

Wall panel junction sealed

Wall panel fixing

Wall tie

Breather membrane

Drained and vented cavity

Cladding

Panel junction

Insulated spline fixed and sealed into end of SIP walls

VCL lapped and sealed at junctions

VCL

Battens forming service void

Internal lining

Figure 2.7 SIP corner

2.7 Airtightness

SIP buildings are able to provide good levels of airtightness due to the panelised construction methodology. A SIP building would normally incorporate a VCL onto the warm side of the external walls. This primarily controls the movement of moisture vapour through the SIP, as well as being an effective air barrier.

The vapour control/air barrier can be lapped and sealed at wall junctions, as well as lapped and sealed at the junctions with other external elements. As with all construction methods, particular attention should be paid to sealing of windows and doors, as well as any service penetrations through the building envelope.

With SIPs, it is also normal practice to seal the splines at panel junctions with either a flexible mastic or expanding foam glue. This, coupled with the vapour control/air barrier, allows SIPs to achieve excellent airtightness performance.

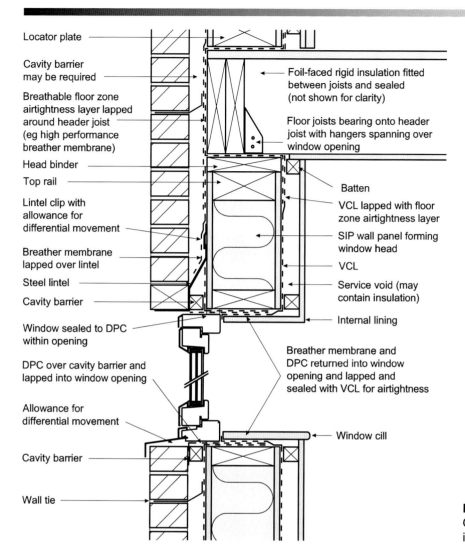

Locator plate

Cavity barrier may be required

Breathable floor zone airtightness layer lapped around header joist (eg high performance breather membrane)

Head binder

Top rail

Lintel clip with allowance for differential movement

Breather membrane lapped over lintel

Steel lintel

Cavity barrier

Window sealed to DPC within opening

DPC over cavity barrier and lapped into window opening

Allowance for differential movement

Cavity barrier

Wall tie

Foil-faced rigid insulation fitted between joists and sealed (not shown for clarity)

Floor joists bearing onto header joist with hangers spanning over window opening

Batten

VCL lapped with floor zone airtightness layer

SIP wall panel forming window head

VCL

Service void (may contain insulation)

Internal lining

Breather membrane and DPC returned into window opening and lapped and sealed with VCL for airtightness

Window cill

Figure 2.8 SIP window
Gaps in construction shown to illustrate membrane laps.

2.8 Condensation risk

Coverings for pitched SIP roofs can be any type that may be used on any other type of roof (such as tiles, slates and profiled metal sheeting). SIP roofs (pitched or flat) are classified as 'cold' roofs, in which the structure of the roof passes through the insulation and is subject to a temperature gradient. Therefore a ventilation void must be provided between the SIP and the roof covering to mitigate the risks of interstitial condensation. The use of counter battens before tiling battens, profiled metal roofing or any other covering should be considered to provide this ventilation. Roofing battens should be fixed to the SIP roof with the use of screws.

National building regulations and British Standards should be consulted for the size of the ventilation voids required in flat and pitched roofs. These regulations and standards also may pose limits on the spans of flat roofs for ventilation requirements.

The guidance given in *BS 5250* regarding ventilation voids and ventilation openings are valid for SIP roof structures and should be followed. In most roof designs, a VCL would also be required to mitigate any risk of interstitial condensation formation within the SIP. This VCL also doubles as an air barrier.

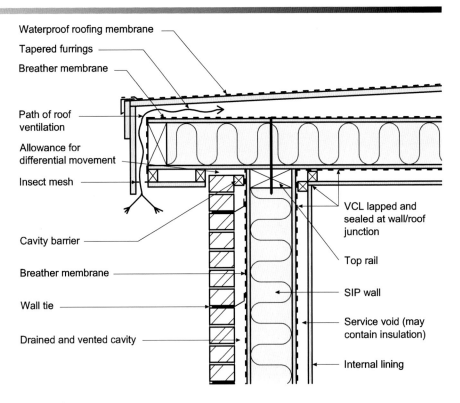

Waterproof roofing membrane

Tapered furrings

Breather membrane

Path of roof ventilation

Allowance for differential movement

Insect mesh

Cavity barrier

Breather membrane

Wall tie

Drained and vented cavity

VCL lapped and sealed at wall/roof junction

Top rail

SIP wall

Service void (may contain insulation)

Internal lining

Figure 2.9 SIP flat roof

2.9 Services

Services should not be installed within the SIP itself as this can seriously affect the structural integrity of the panel. It is normal practice to install a service void, formed with timber battens between the SIP and the wall or ceiling linings, on the inside of the external wall or roof structure. Services can then be run within this void, allowing easy installation and maintenance as well as reducing penetrations through the VCL and air barrier. Where services pass through a SIP, such as pipework for an outside tap or cables for outside lights, the penetration through the SIP must be sealed. The need to de-rate electrical cables and the potential for leaching of plasticiser from the cable must be considered. Where electrical cables pass through a SIP it is prudent to sleeve the cable to avoid these risks.

3 Engineered stud

An engineered stud is a simple way to allow a large depth of insulation to be installed between the loadbearing timber studs used for timber frame wall panels. Engineered I-joists have been in use for many years and the earliest forms of engineered studs were simply these I-joists used vertically. Now a number of different types of engineered stud are available, using either I-joist or metal web joist designs. These studs are available in depths up to 500mm, so they can incorporate as much insulation as is required to achieve the target U-value.

To avoid repetition, the illustrations within this chapter show I-studs, the internal sheathing board providing vapour control and a blown fibre insulation material. Alternative details may show metal web studs, a separate high moisture vapour resistance VCL and quilt/batt/board insulation in any suitable combination.

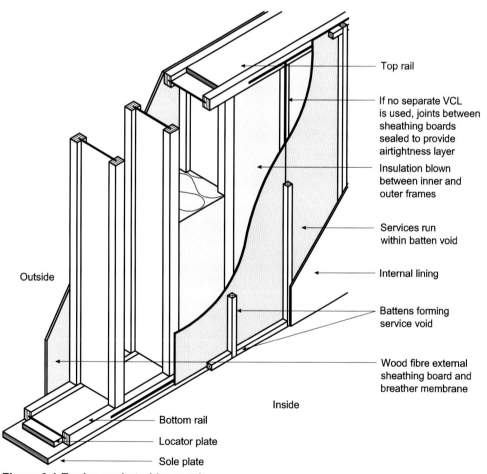

Top rail

If no separate VCL is used, joints between sheathing boards sealed to provide airtightness layer

Insulation blown between inner and outer frames

Services run within batten void

Internal lining

Battens forming service void

Wood fibre external sheathing board and breather membrane

Outside

Inside

Bottom rail

Locator plate

Sole plate

Figure 3.1 Engineered stud isometric

3.1 Applications

Engineered studs can be used as direct replacement for solid timber studs and rails, but provide the user with more space to install insulation and reduce thermal bridging. They can be clad in masonry or lightweight cladding systems, although if a cavity and masonry cladding are added as well, the overall wall thickness can be considerable. Its application is therefore more likely to be with a cladding such as timber or render on board, to keep walls as slender as possible. As well as their use as studs, engineered floor joists can be used in the construction of flat and pitched roofs.

Subject to the approval of a structural engineer, loadbearing engineered stud walls can be used in multi-storey, multi-occupancy buildings.

3.2 Sustainability

As U-values have reduced, standard timber frame studs have become deeper to accommodate thicker insulation. Large-section timber requires larger trees and is therefore a less efficient use of timber than using smaller engineered sections made from young home-grown trees.

3.3 Structure

Engineered studs are normally timber I-joists or open metal web joists used vertically to form studs, although innovations may increase the types of engineered studs commonly available. The design and construction of timber frame walls using engineered timber studs is normally similar to timber frame walls constructed using solid timber studs. The walls consist of vertical loadbearing studs at regular centres, with horizontal top and bottom rails forming prefabricated panels.

Lintels over openings and the cripple studs supporting the lintel are normally constructed using solid engineered timber elements such as glued laminated timber (glulam), laminated veneer lumber (LVL), laminated strand lumber (LSL) and parallel strand lumber (PSL). The use of solid engineered timber around openings also allows for easy and robust fixing of window and door units.

Racking resistance is normally provided with the use of a wood-based board material fixed to either one or both sides of the engineered studs. A common choice is to use OSB on the inside of the wall panels, with a softboard or wood fibre insulation board fixed to the outside of the frame.

Calculation of racking resistance should be conducted in accordance with *Eurocode 5*. The site-specific calculations will determine the specification of board materials and fixings for providing the required racking resistance.

Engineered stud building systems are normally constructed as platform-frame structures (wall panels are room height) with floor structures bearing onto the top of the wall panels. The floor deck then becomes the erection platform for the next storey. They can also be used as balloon-framed panels with intermediate floors supported on ledges fixed to the inside of the walls.

3.4 Manufacture

Engineered stud wall panels are normally constructed off site in a closely controlled factory environment and delivered to site for erection. The design and construction of the wall panels are very similar to standard timber frame construction. Wall panels consist of engineered timber bottom and top rails, with vertical studs and a wood-based sheathing board providing racking resistance.

The panels may be manufactured and delivered as open panels, ready for insulation and linings to be installed on site, or as factory-insulated closed-panel systems. Factory-insulated closed-panel systems must be protected from exposure to moisture during transportation and erection.

Some building warranty providers may require that the structural timber components forming loadbearing timber frame external walls are either naturally durable or preservative treated. Metal web studs can be specified to use preservative-treated timber. The timber components of the studs are preservative treated before the metal webs are pressed into place, so preservative treatment must be specified before the studs are manufactured. Metal web studs cannot be treated after the metal press plates have been installed. Preservative treated timber I-studs are less common, although available from some manufacturers. Timber I-studs cannot be treated with preservative once constructed, so the component parts are treated prior to manufacture. With good building design and due consideration to drainage and ventilation of the timber structure, preservative treatment is normally considered desirable but not essential.

3.5 Erection

As with any other type of timber construction, the wall panels are erected on top of a sole plate. The sole plate is the first piece of timber to be laid on the foundations. Its function is to allow accurate levelling, locating and fixing of the subsequent wall panels. In addition to the sole plate used for metal web studs, most engineered I-stud wall systems use an additional timber called a locator plate. The locator plate is a piece of timber narrower than the sole plate that fits into the void formed by the I-stud bottom rail. This locator plate is fixed down to the sole plate and the wall panel fits over the top, securing the panel in place.

Consideration must be given to how wall panels are to be fixed together. If the wall panels are constructed as open panels (only one side of the wall is sheathed), fixing wall panels together is a relatively simple task. Nails or screws can be driven through the studs and rails of the panel into the adjoining panel or component. Once the structure is fixed, the second layer of sheathing material and insulation can be installed.

If the walls are constructed and delivered as closed panels, site fixing of components becomes more complicated because there is not access to the panel junction. There are a number of possible ways in which closed panel wall structures can be fixed together. These include the use of long screws, timber cover strips at panel junction or proprietary metal fixing systems. A structural engineer will need to determine a site-specific fixing schedule.

3.6 Thermal performance
3.6.1 U-values

The overall depth of the wall and the engineered studs is normally controlled by the target U-value. Typically, engineered studs need to be between 250mm and 500mm deep to give U-values ranging from $0.16\,W/m^2K$ to $0.08\,W/m^2K$ using a blown-in insulation product with a λ-Value of $0.04\,W/mK$.

Although mineral wool and blown insulation products have a high thermal conductivity (less desirable) compared to rigid foam insulation products, they

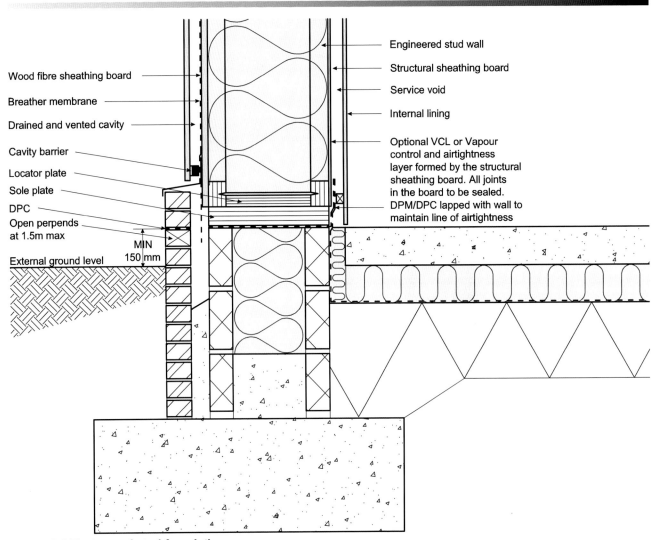

Wood fibre sheathing board

Breather membrane

Drained and vented cavity

Cavity barrier

Locator plate

Sole plate

DPC

Open perpends
at 1.5m max

MIN
150 mm

External ground level

Engineered stud wall

Structural sheathing board

Service void

Internal lining

Optional VCL or Vapour
control and airtightness
layer formed by the structural
sheathing board. All joints
in the board to be sealed.
DPM/DPC lapped with wall to
maintain line of airtightness

Figure 3.2 Engineered stud foundation
Gaps in construction shown to illustrate membrane laps. All structural timber should be at least 150mm above external ground level.

have the advantage of being able to easily fill the voids within an I-stud or metal web stud wall, providing a better continuity of insulation.

U-value calculations normally take into consideration the repeat thermal bridging of the engineered studs, which is much less than that of a solid timber stud. Guidance on the exact calculation of the timber fraction for an engineered timber I-stud can be found in BRE's *BR 443:2006 Conventions for U-value calculations*[5]. If a further reduction in the repeat thermal bridging through the studs is required, additional layers of insulation can be included on either the inside or outside of the structure.

If additional insulation is being installed on the outside of the frame, an insulation product with a low moisture vapour resistance must be chosen to ensure that there is no risk of interstitial condensation at the junction of the timber structure and the insulation. Suitable insulation materials could be high-density mineral wool batts or rigid fibrous insulation products (such as wood fibre insulation boards). Insulation materials used in this type of application should have third-party certification for this use.

L-type stud corner formed using engineered timber studs. Junction detailed with sealant or tape to ensure airtightness is maintained

Insulation blown between wall frames through outer wall leaf. Wood fibre sheathing board and breather membrane to be repaired once insulation is installed.

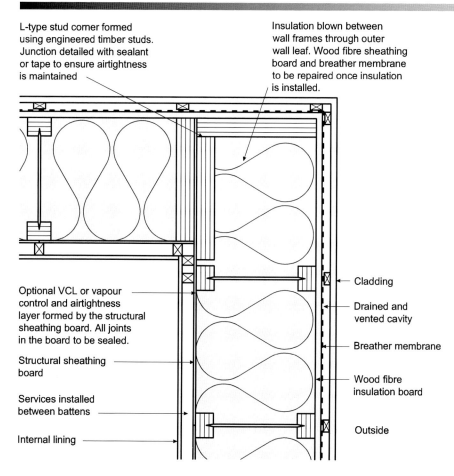

Optional VCL or vapour control and airtightness layer formed by the structural sheathing board. All joints in the board to be sealed.

Structural sheathing board

Services installed between battens

Internal lining

Cladding

Drained and vented cavity

Breather membrane

Wood fibre insulation board

Outside

Figure 3.3 Engineered stud corner

If additional insulation is being installed on the inside of the frame, this could be either a breathable or non-breathable insulation product. Consideration must be given to the location of the VCL in relation to low moisture vapour resistance insulation products and the risk of interstitial condensation.

3.6.2 Ψ-values (thermal bridging)

Engineered stud walls have a greatly reduced cross-section of timber bridging the insulation when compared to solid timber studs, and so the repeat thermal bridging losses through the studs and rails are significantly reduced. *BR 443* contains information about calculating the repeat thermal bridging through engineered studs.

Non-repeat thermal bridges (for example, at building element junctions) are similar in type and location to that of solid-timber systems. Engineered stud walls tend to use a solid engineered timber top and bottom rail, as well as sole plates and head binders. It is these solid engineered timbers that create much of the thermal bridge at junctions, so ways to reduce or eliminate these timbers – and/or the use of additional insulation – should be considered.

3.7 Airtightness

The airtightness performance of engineered stud walls is dependent on the inclusion of an air barrier in the structure. This air barrier could be formed from a number of different elements of the structure.

Depending on the construction of the external wall, there may be a separate high-resistance vapour control membrane present on the inside (warm side)

of the wall. If this is the case, this membrane can also be used as the air barrier. Typically, this type of vapour control and airtightness membrane is polythene or a similar proprietary product. A separate membrane layer allows for simple and robust detailing. Membranes are lapped by at least 100mm, and joints should occur behind service void battens. The batten can be used to compress and seal the joint in the membrane and, in addition, a suitable tape or bead of sealant can also be used to further enhance airtightness performance.

If a separate membrane is not being used (for example, in a wall using OSB sheathing on the inside with a wood fibre or softboard on the outside), then it is typical to use the sheathing board as the airtightness layer. Board materials like OSB and plywood are sufficiently airtight, but the joints and junctions in the board material must be sealed. Joints in the sheathing boards installed in the factory could be sealed to the studs and rails with a flexible sealant or glue. Joints in boards that occur at wall panel junctions could be sealed with either a compressible foam tape installed in the wall panel junction, a flexible gun-applied sealant, or with the use of a proprietary tape. Windows can be sealed into the timber frame using similar methods. Consideration should be given to the difficulties associated with site application of sealant and tapes, such as wet timber, extremes of temperature and dust.

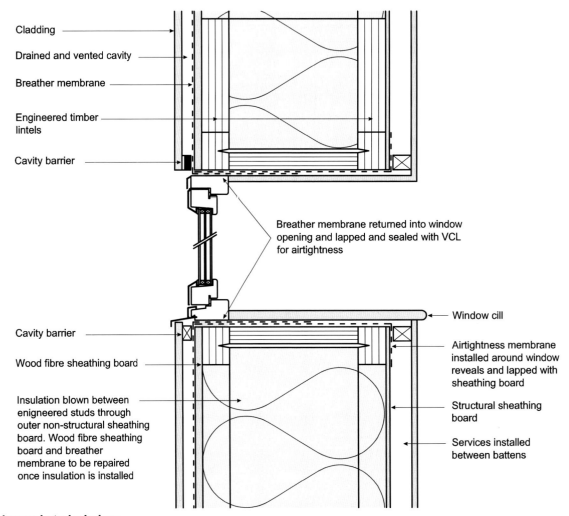

Cladding

Drained and vented cavity

Breather membrane

Engineered timber lintels

Cavity barrier

Breather membrane returned into window opening and lapped and sealed with VCL for airtightness

Window cill

Cavity barrier

Wood fibre sheathing board

Insulation blown between enigneered studs through outer non-structural sheathing board. Wood fibre sheathing board and breather membrane to be repaired once insulation is installed

Airtightness membrane installed around window reveals and lapped with sheathing board

Structural sheathing board

Services installed between battens

Figure 3.4 Engineered stud window
Gaps in construction shown to illustrate membrane laps.

The detailing for airtightness at junctions between elements is reliant on continuity of air barriers and lapping of membranes. At the junction with the ground floor structure, the DPM under a concrete screeded floor can be lapped up the inner face of the timber frame wall and lapped and sealed to either the sheathing or the vapour control and airtightness membrane. This lap or joint with the sheathing is then normally compressed tightly, with a horizontal batten forming the base of the service void.

At the junction with an intermediate floor, continuity of the air barrier on the external wall is interrupted by the floor zone. In order to maintain the line of airtightness, an air barrier membrane can be installed around the end of the floor zone and lapped with the air barrier to the walls above and below. Because the membrane passes over the outside of the floor zone, this air barrier must have a low moisture vapour resistance to ensure that there is no risk of interstitial condensation behind the membrane. At junctions with other elements such as roofs, air barriers should be lapped and sealed to ensure continuity.

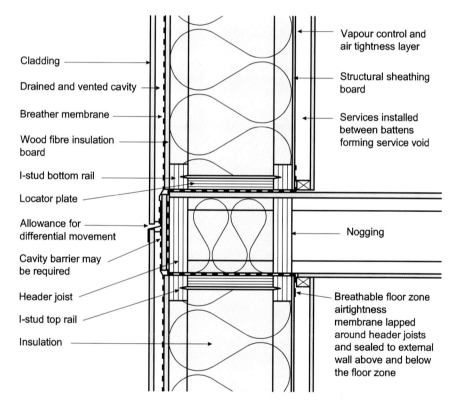

Cladding

Drained and vented cavity

Breather membrane

Wood fibre insulation board

I-stud bottom rail

Locator plate

Allowance for differential movement

Cavity barrier may be required

Header joist

I-stud top rail

Insulation

Vapour control and air tightness layer

Structural sheathing board

Services installed between battens forming service void

Nogging

Breathable floor zone airtightness membrane lapped around header joists and sealed to external wall above and below the floor zone

Figure 3.5 Engineered stud floor
Gaps in construction shown to illustrate membrane laps.

In addition to the overall build envelope airtightness, consideration should always be given to air movement through insulation products, that is, wind washing through fibre insulation products. Wind washing is where air movement through a structure, such as ventilation to roof voids, causes air to pass through an insulation product. The movement of air through the insulation reduces the overall thermal resistance of the insulation, leading to a reduction in performance. As well as air barrier detailing ensuring building envelope airtightness, consideration should also be given to providing an air barrier to prevent air movement through the thermal insulation.

If the outer face of the wall is sheathed with a board material, such as OSB or plywood, this should provide an adequate air barrier. If more permeable

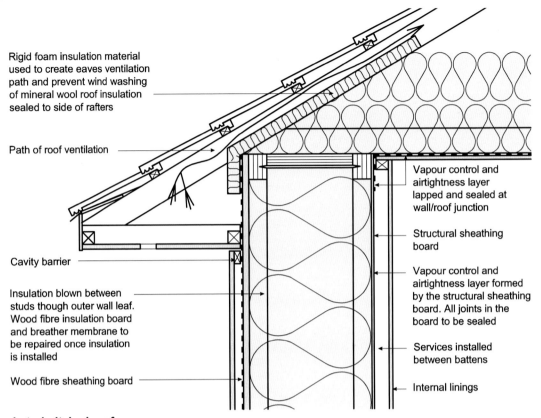

Rigid foam insulation material used to create eaves ventilation path and prevent wind washing of mineral wool roof insulation sealed to side of rafters

Path of roof ventilation

Cavity barrier

Insulation blown between studs though outer wall leaf. Wood fibre insulation board and breather membrane to be repaired once insulation is installed

Wood fibre sheathing board

Vapour control and airtightness layer lapped and sealed at wall/roof junction

Structural sheathing board

Vapour control and airtightness layer formed by the structural sheathing board. All joints in the board to be sealed

Services installed between battens

Internal linings

Figure 3.6 Engineered stud pitched roof

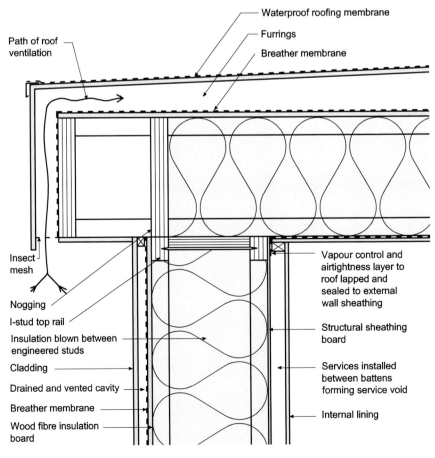

Path of roof ventilation

Waterproof roofing membrane

Furrings

Breather membrane

Insect mesh

Nogging

I-stud top rail

Insulation blown between engineered studs

Cladding

Drained and vented cavity

Breather membrane

Wood fibre insulation board

Vapour control and airtightness layer to roof lapped and sealed to external wall sheathing

Structural sheathing board

Services installed between battens forming service void

Internal lining

Figure 3.7 Engineered stud cold flat roof

materials are used, then the breather membrane on the outside of the wall should be specified, installed and detailed to provide an air barrier function, such as a Type 1 breather membrane as defined in *BS 4016:1997 Specification for flexible building membranes (breather type)*[6], with all joints lapped and sealed with a proprietary tape.

3.8 Condensation risk

To avoid the risk of harmful interstitial condensation, you must consider the moisture vapour resistances of the layers forming the external wall, and include a VCL on the warm side of the insulation (the inside of the wall in the UK climate). As a rule of thumb, the VCL on the warm side of the wall should have a moisture vapour resistance at least five times greater than the layers on the cold side of the wall. If this rule of thumb is followed, no interstitial condensation risk should be present.

In the case of engineered stud walls, the specification of the VCL will be determined by the sheathing material used on the outside of the wall structure. If, for instance, a layer of wood fibre insulation or bitumen-impregnated softboard is installed to the outer face of the studs, a layer of OSB on the inside face of the studs is likely to have a moisture vapour resistance at least five times greater. In this case, the OSB sheathing on the inside of the wall is providing the vapour control function.

If OSB or other board material with a relatively high moisture vapour resistance is installed on the outside of the wall, a high-resistance vapour control membrane may need to be used on the inside of the wall. These high moisture vapour resistance membranes are typically either polythene, or a proprietary vapour control membrane product.

3.9 Services

Typically, a void is formed on the inside of the wall to allow the installation of services. This void is normally formed with vertical timber battens fixed through the sheathing board into the studs. The formation of this void offers a number of benefits, including reduced/eliminated penetration of vapour control and airtightness layer(s) with services, as well as the option to install some additional insulation in the void (if required).

4 Cross-laminated timber

Elsewhere in Europe, solid wood panels have been used in construction for some time. Previously made by stacking and gluing or mechanically fixing full-width sections together, nowadays the thickness of the panel tends to be made up of a number of narrow widths of timber laid together with each layer at right angles to the previous layer. This is known as cross-laminated timber construction (crosslam) and is typical of the solid wood panels now being imported into the UK.

Wall, floor and roof elements can be pre-cut in the factory to any dimension and shape, including openings for doors, windows, stairs, service channels and ducts. The structures offer excellent thermal, acoustic and fire performance to satisfy ever-tougher national building regulations. Crosslam buildings have a very low carbon footprint because carbon absorbed during the trees' growth is locked away inside the wood used to make the panels. Wood is easy to machine, and the material itself is a good insulator. There is no current British or European standard for crosslam, but a standard is currently under development[7].

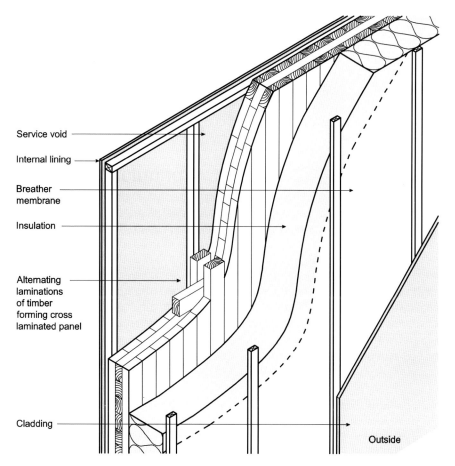

Service void

Internal lining

Breather
membrane

Insulation

Alternating
laminations
of timber
forming cross
laminated panel

Cladding

Outside

Figure 4.1 Crosslam isometric

4.1 Applications

These large solid panels form walls, roofs, floors, and even lift shafts and stairs. The building envelope can be easily clad with other materials such as timber, brick, mineral render or composite panels. Crosslam is now being considered where masonry, concrete and steel have historically been the usual forms of construction. Although crosslam has a promising future in multi-storey construction, it is likely that low-rise non-residential buildings will be its main application. The qualities of exposed crosslam is most appreciated where an internal exposed timber surface offers an aesthetic or acoustic benefit, such as exhibition spaces, schools, places of worship, sports halls, theatres or dwellings.

The benefits of crosslam's off-site manufacture and speed of construction mean it is more likely to be used on large schemes, rather than small one-off projects. The practical and current regulation limit for platform timber frame is 18m to the top-floor deck. In contrast, crosslam is a solid panel, capable of resisting very high racking and vertical loads, and is not limited in height by any building regulation. TRADA Technology has published a scheme design for up to 12 storeys[8]; feasible building heights may increase beyond this in the future.

Crosslam panels have been used in:

- structural and non-structural wall elements
- multi-storey structures with or without concrete substructures
- solid partitions, with and without linings
- floor (ceiling) elements
- parapet wall elements
- roof elements
- pitched roof panels
- cantilevered floors, such as balconies
- curved loadbearing structure
- loadbearing lift shafts
- stairs.

Durability of crosslam panels will depend on the timbers used for its manufacture and the level of their exposure to the weather. Since both sapwood and heartwood are present in crosslam, panels will be liable to decay if their MC exceeds 20% for an extended period of time. Therefore, it is important to protect these timbers from continuous wetting by providing a drained and vented cavity behind the external wall cladding.

4.2 Sustainability

The more wood is used, the nearer the construction moves towards being carbon neutral. Buildings made with solid wood panels in walls, floors and roofs are likely to have a negative carbon footprint. In other words, the carbon absorbed as the trees grow exceeds the carbon associated with all the materials in the building. This is because such large volumes of wood are used in construction. At the end of the building's life, the crosslam panels may be suitable for re-use or recycling. The untreated wood and the glues used in crosslam panels make the product suitable as a biomass fuel.

4.3 Structural

The structural benefits of crosslam are many. Due to the large bearing area available, the walls have high axial load capacity and high shear strength to resist horizontal loads. Lintels are not usually required over openings. The dead weight of crosslam reduces the need for mechanical holding down to resist overturning forces; and buckling in the plane of the wall is unlikely,

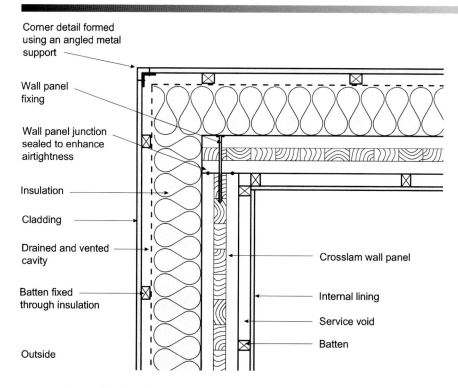

Corner detail formed using an angled metal support

Wall panel fixing

Wall panel junction sealed to enhance airtightness

Insulation

Cladding

Drained and vented cavity

Batten fixed through insulation

Outside

Crosslam wall panel

Internal lining

Service void

Batten

Figure 4.2 Crosslam corner

except for isolated columns and piers. Junctions between panels and elements are fixed using long proprietary screws and/or brackets.

Structural fixings are easy to provide and likely to achieve their design capacity. Second-fix items and cladding can be fixed directly to panels using lightweight power tools.

4.4 Manufacturing

The UK currently imports crosslam panels from Germany, Austria, Switzerland and Sweden. There are at least seven suppliers in the UK. As the UK market develops, a British plant producing crosslam from British timbers may become viable.

Crosslam panels have three, five, seven or more layers, stacked on one another at right angles to each other and glued together in a press over their entire surface area. This cross-laminating and bonding to adjacent pieces results in better dimensional stability, strength in two planes and improved structural integrity. Each layer is composed of softwood boards (of varying lamination thickness) glued together. Sometimes off-cuts are used. The build-up is symmetrical around the middle layer.

The way crosslam is laminated together means that lower-grade timber can be used, as imperfections such as knots are not coincident in adjacent layers. Panel thickness is in the range 50mm to 300mm, but panels as thick as 500mm can be produced. Panels can be up to 20m long and 4.8m wide, but transport constraints normally govern panel size. A maximum length of 13.5m and width of 3m is generally considered practical, but beware of limits dictated by site layout and access. Some suppliers offer curved panels with a minimum radius of 8m.

The manufacturers have various methods of joining adjacent wall and floor panels on site, including rebates, notches and half-lapped joints in the edges of the panels. Panel rebates and openings are cut by CNC routers. Manufacturers can incorporate cut-outs for windows, doors, ducts and chases in the factory. As with all pre-fabricated methods of construction, a design freeze is essential in order to ensure that all openings are correctly incorporated into the panels. Avoid late modifications to openings or additional service runs. Modifying panels on site can be costly and time consuming and may affect their structural integrity.

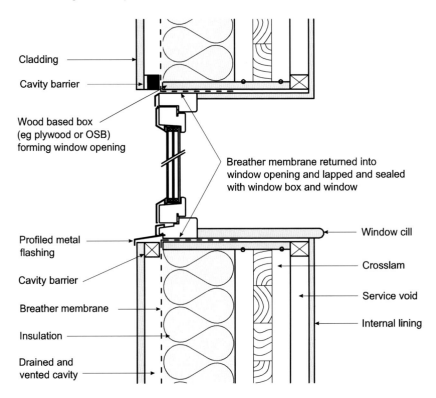

Figure 4.3 Crosslam window
Gaps in construction shown to illustrate membrane laps. Depending on insulation type, cavity barrier may be surface mounted on insulation or pass through insulation back to crosslam wall panel.

Properties vary according to the manufacturer and the basic materials used. The wood is normally spruce, but larch and pine may be available. The common strength grades for the laminates are in the range C16 to C24, and at least one manufacturer offers 'glulam' grades GL24H to GL28H. Moisture content at delivery is typically 8%–14%. Designers will find that working stresses are low due to the large cross-section. Classification of the surface quality of the panels follows *BS EN 13017-1 Solid wood panels. Classification by surface appearance*[9].

Crosslam can be supplied for visible or hidden applications:

- Standard Grade (or non-visible quality): the surface is suitable for lining and typically has top layers corresponding to Class C.
- Interior Grade (residential visible): the surface is suitable for exposed residential internal structure and typically has top layers corresponding to Class AB.
- Interior Grade (industrial visible): the surface that is suitable for exposed industrial internal structure and typically has top layers corresponding to Class BC.

Surfaces are supplied either sanded or planed, depending on manufacturer.

The density of crosslam depends on the timber species and is typically in the range 470 kg/m^3 (spruce) to 590 kg/m^3 (larch) at 12% MC. Crosslam is inherently more stable than the solid wood elements from which it is made. Fissures due to changes in MC are unlikely to appear in Service Class 2 conditions (see *Eurocode 5*).

4.5 Erection

Where site storage is limited, panels can be delivered to site and erected using a 'just-in-time' approach. Provided accuracy in setting out the foundations is achieved, crosslam structures offer reliable on-site programming due to large prefabricated panel elements. Site cutting of panels is not recommended.

Crosslam is suitable for service classes 1 (heated internal) and 2 (unheated internal/covered external) in *Eurocode 5*. Where crosslam is used as an external floor, such as a cantilevered balcony, the top and exposed edges should be protected with a waterproof membrane.

Crosslam panels at ground level should bear onto a treated timber sole plate on a DPC. When crosslam is located adjacent to materials which may transfer moisture to it, they should be separated with a DPC. Structural timber must be located at least 150mm above finished external ground level. If panels are to be exposed in service, take care to avoid water staining and mechanical damage during transport, storage and installation, as blemishes can be difficult to rectify.

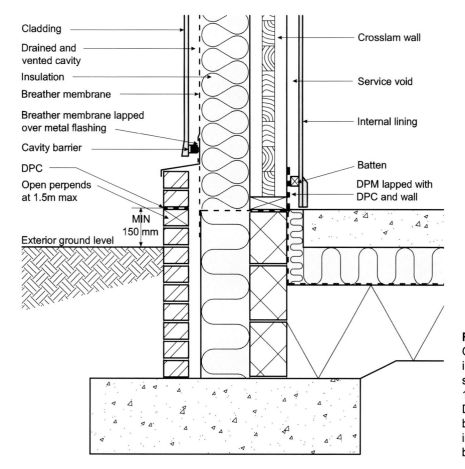

Figure 4.4 Crosslam foundation
Gaps in construction shown to illustrate membrane laps. All structural timber should be at least 150mm above external ground level. Depending on insulation type, cavity barrier may be surface mounted on insulation or pass through insulation back to crosslam wall panel.

Connections between panels are typically screw-fixed using lightweight power tools. The relatively low level of noise and disruption on a crosslam site offers advantages over in-fill sites, where the impact on neighbours is an important consideration. While installation in wet weather is inconvenient, rain has no immediate effect on the panels as they will dry quickly. Temporary protection for panels that are left exposed to weather for a number of days is recommended.

Studies have shown that crosslam buildings are subject to small amounts of differential movement, approximately 3mm–5mm per storey height. Details for cladding, vertical services and linings spanning floor zones should allow for this movement.

Where a roof has a simple arrangement of continuous ridges and gable ends, a room in the roof can be formed using crosslam panels. A breathable roof underlay and rigid insulation will normally be located above the crosslam panels to give a 'warm roof' construction. Crosslam pitched roof panels are typically supported by external walls or purlins and ridge beams. The crosslam panels must transfer vertical load to the wall plates and purlins and ridge beams using suitably engineered fixings. The ridge beam can be omitted where there is sufficient provision to resist the horizontal thrust of the roof at the eaves using a thrust plate and floor diaphragm.

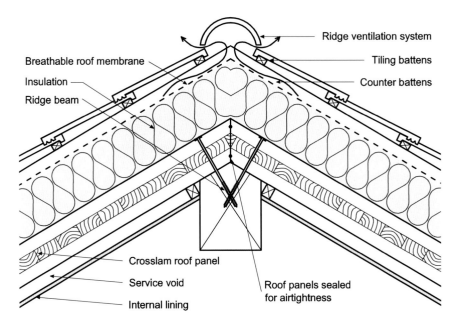

Figure 4.5 Crosslam roof ridge

4.6 Thermal performance

Unlike conventional timber framing alone, crosslam makes a contribution to the U-value. For example, a 100mm crosslam panel of density $500\,\text{kg/m}^3$ has λ value $0.13\,\text{W/mK}$. Therefore to achieve the target U-value $0.35\,\text{W/m}^2\text{K}$ under the current Building Regulations for England and Wales, a 100mm crosslam panel (with 12mm plasterboard, 25mm service void, 50mm cavity, brick cladding and no thermal bridging) would require 75mm mineral wool insulation or equivalent.

Thermal mass can reduce the variation in temperature over the daily cycle, although other factors such as ventilation, solar gain and insulation must be

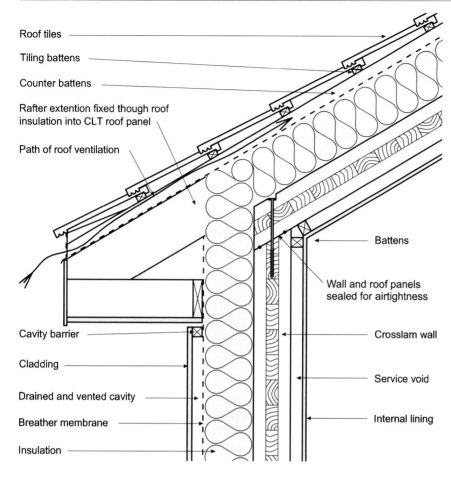

Roof tiles

Tiling battens

Counter battens

Rafter extention fixed though roof insulation into CLT roof panel

Path of roof ventilation

Battens

Wall and roof panels sealed for airtightness

Cavity barrier

Cladding

Crosslam wall

Drained and vented cavity

Service void

Breather membrane

Internal lining

Insulation

Figure 4.6 Crosslam pitched roof
Gaps in construction shown to illustrate membrane laps. Depending on insulation type, cavity barrier may be surface mounted on insulation or pass through insulation back to crosslam wall panel.

taken into account. When the design maximises passive solar gain through glazing or from heat generated in the building's use, the thermal mass of the crosslam can be used to collect and store energy during the day for emission later in the cycle. Important properties are thermal conductivity (rate of transferring heat) and specific heat capacity (ability to retain heat). For example, crosslam and lightweight concrete block materials have a similar thermal conductivity, while crosslam has a greater specific heat capacity. Therefore, a 70mm crosslam panel has a thermal mass similar to 100mm lightweight block.

Crosslam and platform-frame timber can be combined to produce a more efficient structural form. In this form, crosslam floor panels span parallel to external walls so that external walls can be highly insulated non-loadbearing in-fill panels. Care must be taken with potential differential movement between the loadbearing and in-fill panels.

It may also be possible to use crosslam panels as pre-insulated wall and roof cassettes, but care would be needed to avoid damaging the insulation in transit.

4.7 Condensation risk

Crosslam is vapour permeable. Designers must assess the risk of condensation. A well-ventilated cavity will be needed. Roofs should be designed and constructed as warm structures, with all insulation and waterproof membranes located above the crosslam.

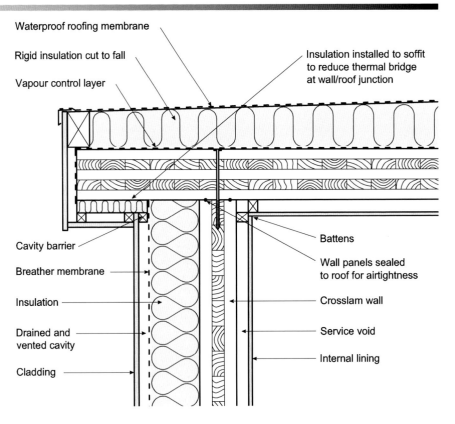

Waterproof roofing membrane

Rigid insulation cut to fall

Vapour control layer

Insulation installed to soffit
to reduce thermal bridge
at wall/roof junction

Cavity barrier

Breather membrane

Insulation

Drained and
vented cavity

Cladding

Battens

Wall panels sealed
to roof for airtightness

Crosslam wall

Service void

Internal lining

Figure 4.7 Crosslam flat roof

4.8 Airtightness

Thermal performance will be compromised if the construction does not achieve adequate airtightness. Because crosslam construction would not normally include a VCL, the system relies entirely on the detailing of joints

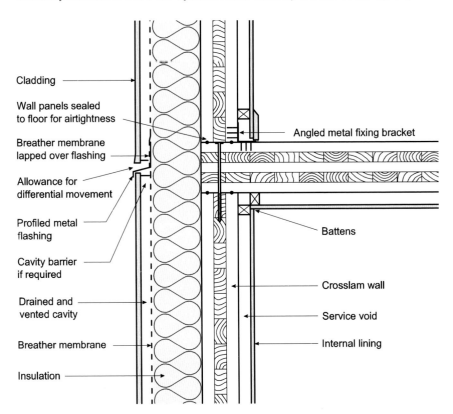

Cladding

Wall panels sealed
to floor for airtightness

Breather membrane
lapped over flashing

Allowance for
differential movement

Profiled metal
flashing

Cavity barrier
if required

Drained and
vented cavity

Breather membrane

Insulation

Angled metal fixing bracket

Battens

Crosslam wall

Service void

Internal lining

Figure 4.8 Crosslam intermediate floor
Gaps in construction shown to illustrate membrane laps. Depending on insulation type, cavity barrier may be surface mounted on insulation or pass through insulation back to crosslam wall panel.

to achieve airtightness. Joints that are merely screwed together may suffice, but this will depend on 'true' surfaces, good workmanship and the level of airtightness specified. Airtightness is normally achieved with either pre-compressed foam tape within the joint and/or tape across the joint. Pre-completion testing may be needed to demonstrate compliance.

4.9 Services

Panels can incorporate pre-formed service routes for small services such as electrical wiring. These may be routed through the rebates at the interfaces between two adjacent panels or be drilled and rebated into the panels to order. Services may also be routed in chases on the surface of the panels which are milled into the panels in the factory. Early coordination with the architect and building services engineer is essential to take advantage of this service. Alternatively, building services may be routed in a cavity formed by a layer of plasterboard fixed to battens on the inner face of the crosslam wall, or a suspended ceiling supported from the underside of floor panels.

Fixing decorations and fittings to plain crosslam walls is easier to achieve than with timber frame, concrete or masonry walls. There is no need to identify stud locations, or install wall plugs before installing fixings.

Under floor heating pipework or other services that run above the floor panel can be accommodated within a screed.

5 Twin stud walls

A twin stud wall is a simple, easy method of getting a lot of insulation into a wall with relatively inexpensive materials. It is simply two timber frame stud walls in parallel, separated by a cavity, but only one of these carries the vertical load of the building. Depending on the structural design and order of work proposed on site, this can be either the inner or the outer wall. The cavity width and the depth of both wall studs can be varied, to provide sufficient timber for structural performance and space for insulation. This simple method uses many materials and techniques familiar to those who build standard timber frame buildings.

To avoid repetition, the illustrations within this chapter show internal loadbearing studs, a blown fibre insulation material and the internal sheathing board providing racking resistance. Alternative details may show external loadbearing studs and sheathing, or quilt/batt/board insulation in any

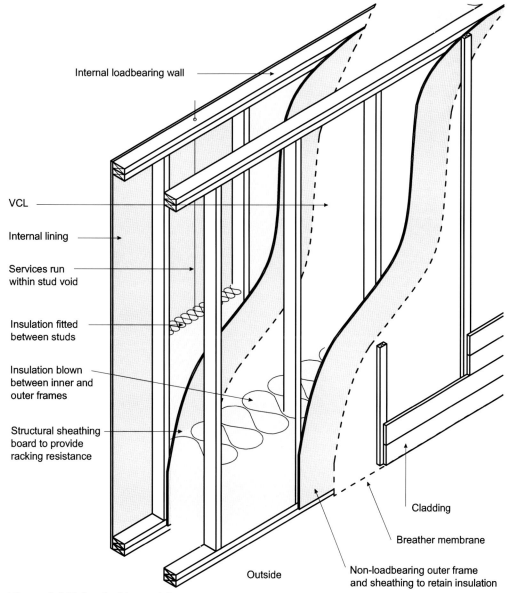

Internal loadbearing wall

VCL

Internal lining

Services run
within stud void

Insulation fitted
between studs

Insulation blown
between inner and
outer frames

Structural sheathing
board to provide
racking resistance

Cladding

Breather membrane

Non-loadbearing outer frame
and sheathing to retain insulation

Outside

Figure 5.1 Twin stud isometric

suitable combination. The drawings indicate an external wall depth in excess of 400mm to avoid risk of interstitial condensation on the VCL.

5.1 Applications

In timber frame construction, the twin stud method has been used for many years in party walls for acoustic separation. In some parts of the world, it has also been used in external walls; and with requirements for improved thermal performance, it is now being used in the UK too. The additional stud and cavity used in this external wall construction adds to the overall wall thickness. If a cavity and masonry cladding are added too, the overall wall thickness can be considerable. Its application is therefore more likely to be with a cladding such as timber or render on board, to keep walls as slender as possible. On sites where wall thickness is not so important, brick cladding may be an equal consideration. Floor and roof designs are the same as other timber frame building techniques and are not examined here.

With energy prices rising and legislation pushing for the improvement of our existing housing stock, twin stud wall design could be used to improve current timber frame buildings. Adding an extra wall leaf externally is preferred, as this reduces disturbance to occupants and provides a continuous thermal layer over floor zones and lintels. It also reduces thermal bridging and reduces the risk of interstitial condensation becoming an issue with the upgrading of the thermal performance. The existing external cladding can be removed and an additional timber frame wall can be built outside the current structure before the wall is insulated and then reclad. If external space is not available, an extra wall leaf could be constructed internally. Internal linings, services, and possibly the VCL would need to be removed and refixed along with additional insulation once the extra wall leaf had been constructed.

5.2 Sustainability

As U-values have reduced, standard timber frame studs have become deeper to accommodate thicker insulation. Large-section timber requires larger trees and is therefore less efficient use of timber than using smaller sections. In twin stud walls, studs do not run through the full thickness of the wall, so smaller-section timber can be used to provide the structural performance and support for cladding/linings.

Once a twin stud wall has been chosen, very low U-values can be achieved at minimal extra cost. The extra cost of the cavity insulation and the sacrifice of space to the cavity are the only barriers to achieving a superior external wall with negligible heat loss. The long-term energy and cost savings benefits of such a wall must be considered.

5.3 Structural

Only one leaf of a twin stud wall will be designed to be loadbearing. Whether this is internal or external will affect the overall design of the wall panels and floor joists. There are advantages and disadvantages to both methods, which have an impact on the design and construction of the whole structure.

5.3.1 Outer wall loadbearing

This method is similar to modern platform timber frame. The outer timber frame leaf is loadbearing, taking vertical load from the external walls, roof and floor zones above, as well as lateral load from internal loadbearing walls and external claddings. The inner wall leaf is designed to be non-loadbearing and is constructed to fit around these elements. It provides a frame which will accommodate insulation, a VCL, services and fixing material for wall linings.

5.3.2 Inner wall loadbearing

With the inner wall leaf loadbearing, the outer wall should be self supporting from its own sole plate. The outer timber frame wall leaf becomes a continuous vertical element with a cavity on both sides of it. However, it must be designed and constructed to provide adequate restraint against wind load and cladding weight. In North America, it has become popular with twin stud wall construction to cantilever the outer non-loadbearing wall from the base of the loadbearing inner leaf and thereby remove the need to provide additional foundations for it. Due to the loadings, wall height would be limited using a cantilevered method.

Whichever method is chosen, it is important to remember that while differential movement will occur, the timber frame should be designed and constructed so that movement is equal in each wall leaf. This can be achieved with tight construction and by ensuring equal amounts of

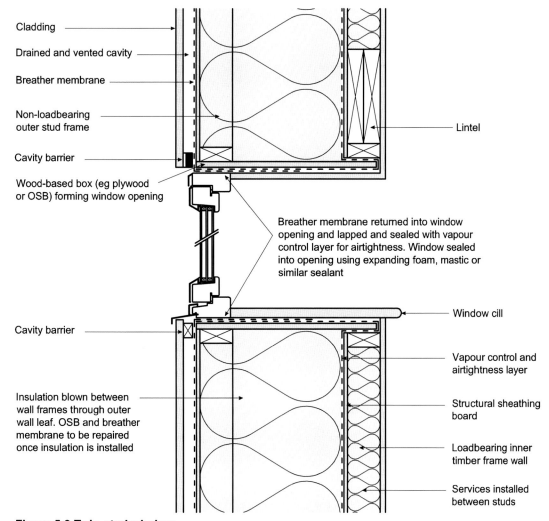

Figure 5.2 Twin stud window
Gaps in construction shown to illustrate membrane laps.

horizontal cross-grain timber in each wall. If the twin frames do not move downwards together, damage will occur where they are connected together, such as around openings.

5.4 Manufacturing

The materials and techniques used to manufacture twin stud walls are the same as for conventional platform timber frame. This includes the need to treat timber with preservative if required.

As its name suggests, a twin stud wall requires double the number of external wall panels to be designed, manufactured and delivered to site. Beyond the need to design wall panels for structural performance, placement of studs must take into account wall lining edge support on the internal leaf and support for cladding fixings on the outer leaf. The non-structural wall leaf could have fewer studs at wider centres than the structural leaf. Accurate placement of studs around openings is essential to ensure they 'line through' when assembled on site.

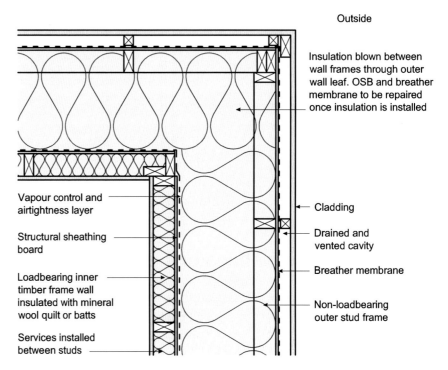

Outside

Insulation blown between wall frames through outer wall leaf. OSB and breather membrane to be repaired once insulation is installed

Vapour control and airtightness layer

Structural sheathing board

Loadbearing inner timber frame wall insulated with mineral wool quilt or batts

Services installed between studs

Cladding

Drained and vented cavity

Breather membrane

Non-loadbearing outer stud frame

Figure 5.3 Twin stud corner
Loadbearing inner frame. Outer frame erected after the VCL has been applied to the outside of the inner structural frame. Gaps in construction shown to illustrate membrane laps.

To provide racking resistance to the structure and/or to protect the integrity of the polythene vapour control/air barrier, it is typical to provide a timber-based sheathing board on the outer face of the inner wall. If the outer wall leaf is structural, this too will be sheathed.

The alternative to building wall panels in a factory is the on-site 'stick building' method. It has a high skilled-labour element but low material costs, so lends itself well to the self-build market. Stick build is the predominant method of timber frame construction in North America.

5.5 Erection

Conventional platform-frame techniques can be used to construct much of the timber frame. Accuracy in setting out is essential to ensure all panels fit together as designed. The loadbearing walls must be erected first, whether they are inner or outer. The outer face of the inner wall is the typical location for the VCL to be installed. This can create order-of-work issues when installing insulation, as slab types of insulation will need to be installed before the timber frame is weather tight. One method of ensuring that insulation is installed dry and remains so is to use blown insulation, which is pumped into the wall cavity and stud voids following erection of the timber frame and completion of a weathertight external envelope.

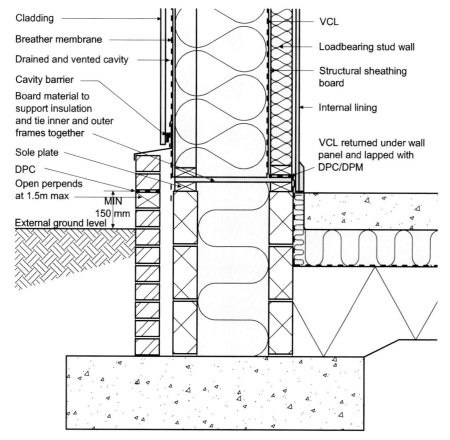

Cladding
Breather membrane
Drained and vented cavity
Cavity barrier
Board material to support insulation and tie inner and outer frames together
Sole plate
DPC
Open perpends at 1.5m max
MIN 150mm
External ground level

VCL
Loadbearing stud wall
Structural sheathing board
Internal lining
VCL returned under wall panel and lapped with DPC/DPM

Figure 5.4 Twin stud foundation
Gaps in construction shown to illustrate membrane laps. All structural timber should be at least 150mm above external ground level.

If the inner leaf of the external wall is non-loadbearing, it may be erected after the loadbearing outer structure is complete. Pre-manufactured walls would need to be stored inside the building, close to their intended location, while access is available. They would also need to be manufactured slightly shorter than the loadbearing walls to allow them to be lifted into place.

5.6 Thermal performance

The dimensions of the gap between the studs and the stud section are dependent on the thermal performance and structural requirements of the building. U-value performance is therefore entirely variable and can precisely meet any given specification. Best practice for reduced thermal bridging is to build the two walls so that the studs are offset, allowing the maximum amount of insulation at all locations. This is not practical around openings, and requires diligence from the timber frame designer to achieve elsewhere.

Thermal bridging does exist at connections between the twin wall leafs required for lateral restraint. These are typically limited to connections at the

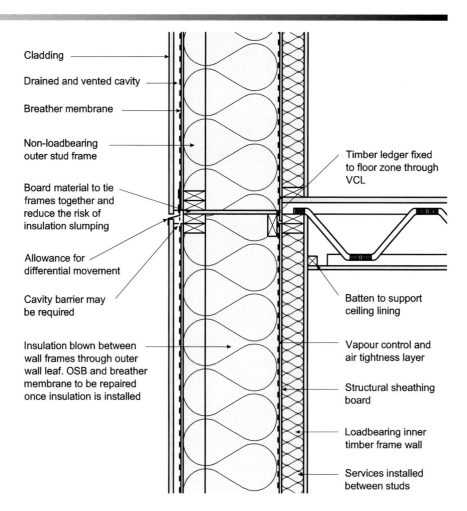

Cladding

Drained and vented cavity

Breather membrane

Non-loadbearing
outer stud frame

Board material to tie
frames together and
reduce the risk of
insulation slumping

Allowance for
differential movement

Cavity barrier may
be required

Insulation blown between
wall frames through outer
wall leaf. OSB and breather
membrane to be repaired
once insulation is installed

Timber ledger fixed
to floor zone through
VCL

Batten to support
ceiling lining

Vapour control and
air tightness layer

Structural sheathing
board

Loadbearing inner
timber frame wall

Services installed
between studs

Figure 5.5 Twin stud intermediate floor zone
Gaps in construction shown to illustrate membrane laps.

floor zones at each storey height, plus around openings and service penetrations.

5.7 Airtightness and condensation

When the internal leaf of a twin stud wall is loadbearing, the airtightness/ vapour control layer can be installed on the outer face of the internal leaf once it is erected. This can cover both the walls and edge of the floor zone to form a continuous membrane. It can then provide temporary weather protection to the building, while the outer leaf is erected. Lateral support for the external non-loadbearing wall leaf can be provided at floor zones, either with fixings sealed through the membrane, or by fixing a ledger to the outer edge of the floor zone and fixing to that, once the membrane is in place.

There needs to be a sufficient quantity of insulation on the outer face of the VCL versus the inner face of the VCL. This ensures that there is no condensation risk within the wall structure, by keeping the temperature of the membrane above the dew point. By avoiding penetration damage to the vapour control/air barrier, both airtightness and vapour control are improved. However, when the inner wall leaf is non-loadbearing, placing an effective membrane in this location is impractical.

When the inner leaf of a twin stud wall is non-loadbearing, the routing of the vapour control/air barrier is more complicated and has to take into consideration the order of work on site. One option is to install it on the inner face of the non-loadbearing wall and then create a battened service zone inside it. This further adds to the overall wall thickness. Using this method, vapour control/air barriers must link through floor zones between and around joists, or be elongated around the ends of the joists to the external wall and back again (using a breathable air barrier to avoid interstitial condensation).

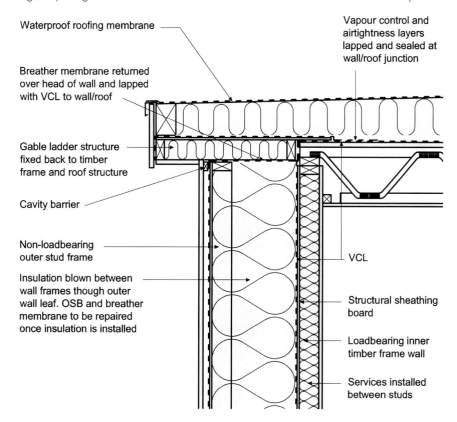

Waterproof roofing membrane

Breather membrane returned over head of wall and lapped with VCL to wall/roof

Gable ladder structure fixed back to timber frame and roof structure

Cavity barrier

Non-loadbearing outer stud frame

Insulation blown between wall frames though outer wall leaf. OSB and breather membrane to be repaired once insulation is installed

Vapour control and airtightness layers lapped and sealed at wall/roof junction

VCL

Structural sheathing board

Loadbearing inner timber frame wall

Services installed between studs

Figure 5.6 Twin stud flat roof

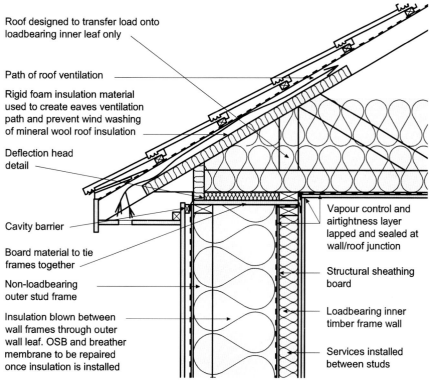

Roof designed to transfer load onto loadbearing inner leaf only

Path of roof ventilation

Rigid foam insulation material used to create eaves ventilation path and prevent wind washing of mineral wool roof insulation

Deflection head detail

Cavity barrier

Board material to tie frames together

Non-loadbearing outer stud frame

Insulation blown between wall frames through outer wall leaf. OSB and breather membrane to be repaired once insulation is installed

Vapour control and airtightness layer lapped and sealed at wall/roof junction

Structural sheathing board

Loadbearing inner timber frame wall

Services installed between studs

Figure 5.7 Twin stud pitched roof
Gaps in construction shown to illustrate membrane laps.

Although vapour control must be provided on the warm side of the insulation, a more effective air barrier location is on the outer face of the outer loadbearing leaf, using a breathable air barrier/breather membrane. External air barriers are not a common detail in the UK and require careful sealing at junctions and perimeters to be effective.

5.8 Services

The routing of services in twin stud construction follows a similar procedure to standard timber frame. Depending on the location of the vapour control/air barrier, electrical and plumbing services installed in external walls will either be located within the inner leaf, or in a separate internal service zone. By avoiding penetration damage to the vapour control/air barrier, both airtightness and vapour control are improved.

6 Looking ahead

As the construction industry moves towards its zero-carbon goal, the innovative timber construction forms described in this book are likely to become more commonplace in the UK. Meanwhile, there are other innovations emerging in the community of 'enthusiast' builders. When the more robust alternatives become 'mainstream', these could be included in future editions.

The traditional platform timber frame form described in *Timber frame construction*, which we have become used to over the last 50 years, will continue to evolve. In the short to medium term, 'standard' timber frame utilising solid timber studs, 140mm in depth, will still be the most common choice of timber construction due to its convenience, familiarity and a proven track record of good performance. As *Timber frame construction* shows, excellent U-values can be achieved with the use of additional insulation layers over the inner or outer face of the timber structure.

New materials, methods of working, regulations, time and cost will all play their part in the evolution of the next generation of timber frame. What is becoming clear is that the term 'timber frame construction' will only loosely define this method, with numerous hybrid forms and variants competing for ease of construction, performance and cost.

We hope this book will enlighten and inspire the timber construction industry to build using innovative timber forms and methods. These are still in their infancy, some indeed a minority interest, so the information provided here is a starting point for further development. Whatever form or variant is used, it is important to follow the advice in this book (as well as *Timber frame construction*) regarding designing for durability. Ensuring that the timber structure remains dry and is allowed to drain and ventilate if it does get wet are vital to long-term durability.

It is encouraging that in recent years there has been a significant increase in the use and public profile of innovative timber structures. Many of these projects have used innovative construction forms, predominantly crosslam and engineered studs, to deliver exciting, energy-efficient, low-carbon buildings. These buildings are pointing to a bright future for the use of timber in construction.

The demand for low-carbon, low-energy buildings is likely to grow rapidly over the coming years. The innovative use of timber in the construction of these buildings will help to achieve these goals. It is these innovations in timber that we hope will grow, prosper and lead to a more sustainable, timber-orientated construction industry in the future.

References

1 Lancashire, R. and Taylor, L., Timber frame construction, 5th edition, ISBN 978-1900510820, TRADA Technology, 2011

2 BS 5250:2002 Code of practice for control of condensation in buildings, BSI

3 BS EN 1995-1-1:2004+A1:2008 Eurocode 5. Design of timber structures. General. Common rules and rules for buildings, BSI

4 BS EN ISO 10211-1:1996 Thermal bridges in building construction. Heat flows and surface temperatures. General calculation methods, BSI

5 BR 443:Conventions for U-value calculations, ISBN 1860819249, BRE, 2006

6 BS 4016:1997 Specification for flexible building membranes (breather type), BSI

7 O/WD 16696 Timber structures – Cross laminated timber – Component performance and production requirements, (under development in July 2012)

8 Worked example: 12-storey building of cross-laminated timber (Eurocode 5), TRADA Technology, 2009

9 BS EN 13017-1:2001 Solid wood panels. Classification by surface appearance. Solid wood panels. Classification by surface appearance. Softwood, BSI

Further reading

England and Wales Building Regulations: Approved Document A (Structure), NBS, 2010, available at www.planningportal.gov.uk

England and Wales Building Regulations: Approved Document B (Fire Safety), Volumes 1 and 2, Dwellinghouses and Buildings other than dwellinghouses, NBS, 2010, available at www.planningportal.gov.uk

England and Wales Building Regulations: Approved Document E (Resistance to the passage of sound), NBS, 2004, available at www.planningportal.gov.uk

Eurocode 5: timber design essentials for engineers, ISBN 978-1900510707, TRADA Technology, 2009

GD10: Cross-laminated timber (Eurocode 5) design guide for project feasibility, ISBN 9781900510691, TRADA Technology, 2009

Hairstans, R., Off-site and modern methods of timber construction: a sustainable approach, ISBN 978-1900510738, TRADA Technology Ltd, 2010

Pitts, G. and Lancashire R., Low energy timber frame buildings: designing for high performance, 2nd edition, ISBN 978-1900510806, TRADA Technology Ltd, 2011

Technical Booklets (Northern Ireland Building Regulations), available at www.dfpni.gov.uk

Technical Handbooks: Domestic, October 2011, available at www.scotland.gov.uk/publications

Technical Handbooks: Non domestic, October 2011, available at www.scotland.gov.uk/publications

Timber and the Sustainable Home: Eight architects debate the challenges ahead, ISBN 978-1900510578, TRADA Technology Ltd, 2008

WIS 2/3-61: Cross-laminated timber: Introduction for specifiers, TRADA Technology, 2011

WIS 2/3-62: Cross-laminated timber: Structural principles, TRADA Technology, 2011

TRADA bookshop

Timber frame construction 5th edition

Robin Lancashire, Lewis Taylor

Timber frame construction is the definitive design guide for timber frame buildings:

- Incorporates latest requirements for Part L and increased performance.

- All drawings were revised.

- Guidance for both Eurocode 5 and BS 5268 are now included.

- Recommended by UK Timber Frame Association

ISBN 9781900510820 264pp. 2011.
£60.00 (TRADA members £35.00).
Available format: Paperback.

Off-site and modern methods of timber construction. A sustainable approach.

Robert Hairstans

An authoritative guide for designers, construction professionals and manufacturers. Links timber's traditional qualities and the latest production technologies, an inspiration for many modern, appealing buildings.

ISBN 9781900510738. 96pp. 2010.
£35.00 (TRADA Members £25.00).
Available format: Paperback.

Low energy timber frame buildings. Designing for high performance.

TRADA Technology 2011
Geoffrey C Pitts. Advisory editor: Robin Lancashire.

This illustrated guide demonstrates how timber frame buildings can be designed and constructed cost effectively to be energy efficient. It shows how designers can also play a role in building use, providing end-users with a comprehensive operation manual. The design principles and construction details described apply to most mainstream building types up to seven storeys high.

ISBN 9781900510806, 80pp. 2011.
£45.00 (TRADA Members £25.00).
Available format: Paperback.

The site manager's guide to timber frame construction

Lewis Taylor, Robin Lancashire

This handy illustrated pocket guide concentrates on the most important aspects of assembly of the timber frame and the ancillary operations involved in completing the superstructure of the building. Typical tasks undertaken by different members of the construction team are included.

ISBN 9781900510653. 64pp. 2009.
£25.00 (TRADA members £19.50).
Available format: Paperback.

Timber in contemporary architecture

Peter Ross, Giles Downes & Andrew Lawrence
A designer's guide

Written by leading experts in timber architecture and engineering, this TRADA 75th anniversary publication investigates materials, connections, applications, and celebrates innovation. Excellence in timber design is demonstrated in the last section of the book which is devoted to eighteen highly illustrated case studies.

ISBN: 9781900510660. 2009.
£55.00 (TRADA members £40.00).
Available format: Hardback.

Essential timber frame standard details

Saving countless hours of detail design, Essential Timber Frame Standard Details (for single occupancy detached, semi-detached and terraced houses) contains 32 of the most commonly-used drawings from TRADA Technology's book and CD Timber frame: standard details for houses and flats in a ready-to-use electronic format.

Downloadable application. 2010.
£190.00 + VAT (TRADA members £95.00 + VAT).

TRADA bookshop

Eurocode 5 span tables for solid timber members in floors, ceilings and roofs for dwellings

Now revised and updated to comply with Eurocode 5, new features in this 3rd edition include: Timber specifications to BS EN 1912, loads from Eurocode 1 including the National Annex, plus new snow loads for the UK. The calculations apply to buildings up to three storeys in height above ground level. Span tables were originally included in the England and Wales Building Regulations, Approved Document A: Structure and subsequently revised by TRADA.

ISBN: 9781900510714. 56pp. 2009.
£23.00 (TRADA members £20.00).
Available format: Paperback.

Concise illustrated guide to timber connections

Illustrated guide to timber connections

- brings together architectural and structural considerations

- illustrated with connection details and assembly forms

- researched and written by TRADA Technology, the experts in timber construction

This concise illustrated guide to timber connections aims to help architects and engineers answer four questions:

- What kind of connection is needed?

- What might it look like?

- How will the structural constraints affect its configuration and appearance?

- What are the best details for durability and long-lasting visual appeal?

ISBN 9781900510851. 32pp. 2012.
PDF free to TRADA Members.
Printed copy £19.50 (TRADA Members) and £30.00 (non members).

TimberWISE CD ROM

Contains 69 electronic Wood Information Sheets, accessible in a portable, searchable format. Latest TRADA Regulatory Construction Briefings and Technical Information Sheets from Chiltern International Fire are included. The timberWISE price banding has been restructured to better represent our customer's usage. The current rates (for TRADA members) reflect the size of an organization in line with our membership subscription rates.

Annual subscription rates

TRADA Members	

Specifier / Professional Members	
Sole practitioner, small contractor	£65.00+VAT
Medium, large practices, large contractors	£85.00+VAT

Timber Industry/Corporate Members	
Bands A-C	£85.00+VAT
Bands D-H	£150.00+VAT

Non Members	£175.00

NEW LOWER PRICE *Non-members

Eurocode 5: timber design essentials for engineers

The objective of this publication is to pull together TRADA's engineering guidance documents on Eurocode 5 (EC5) into one, easy-to-use reference. Includes sections on limiting vibrations, material properties, designing flitch beams, multiple fasteners, with cross-laminated timber, and more.

2009.
£174 + VAT (TRADA members £90.00 + VAT).
Available format: PDF only.

Order all publications at www.trada.co.uk/bookshop or by calling 01494 569602

56